4TH EDITION
Dilapidations

RICS guidance note

Please note: Reference to the masculine include, where appropriate, the feminine.

Published by RICS Business Services Limited
a wholly owned subsidiary of
The Royal Institution of Chartered Surveyors
under the RICS Books imprint
Surveyor Court
Westwood Business Park
Coventry CV4 8JE
UK

No responsibility for loss occasioned to any person acting or refraining from action as a result of the material included in this publication can be accepted by the author or publisher.

ISBN 1 84219 075 X

© RICS June 2003. Copyright in all or part of this publication rests with the RICS, and save by prior consent of the RICS, no part or parts shall be reproduced or otherwise, now known or to be devised.

Typeset in Great Britain by Wyvern 21 Ltd, Bristol, UK.
Printed in Great Britain by MFK Ltd, Stevenage, UK.

Contents

RICS Guidance Notes	4

    1 Introduction
    2 Taking Instructions
    3 Role of the Surveyor
    4 Information
    5 The Schedule
    6 The Claim
    7 Dialogue

Appendix 1	**The Protocol and Annex**	19
	Property Litigation Association Protocol for Terminal Dilapidations Claims for Damages	20
	Annex Schedule of Dilapidations	24
Appendix 2	**Schedule of Other References**	27
Appendix 3	**Example of a Schedule of Dilapidations**	29
Appendix 4	**Recommended Form of Scott Schedule**	33
Appendix 5	**Example of Scott Schedule**	39
Appendix 6	**Bibliography**	43
Appendix 7	**Statutory Material**	45
	Law of Property Act 1925	47
	Landlord and Tenant Act 1927	50
	Leasehold Property (Repairs) Act 1938	51
	Landlord and Tenant Act 1954	54

RICS Guidance Notes

This is a guidance note. It provides advice to members of the RICS on aspects of the profession. Where procedures are recommended for specific professional tasks, these are intended to embody 'best practice', i.e. procedures which in the opinion of the RICS meet a high standard of professional competence.

Members are not required to follow the advice and recommendations contained in the note. They should however note the following points.
When an allegation of professional negligence is made against a surveyor, the court is likely to take account of the contents of any relevant guidance notes published by the RICS in deciding whether or not the surveyor had acted with reasonable competence.

In the opinion of the RICS, a member conforming to the practices recommended in this note should have at least a partial defence to an allegation of negligence by virtue of having followed those practices. However, members have the responsibility of deciding when it is appropriate to follow the guidance. If it is followed in an appropriate case, the member will not be exonerated merely because the recommendations were found in an RICS guidance note.

On the other hand, it does not follow that a member will be adjudged negligent if he has not followed the practices recommended in this note. It is for each individual surveyor to decide on the appropriate procedure to follow in any professional task. However, where members depart from the good practice recommended in this note, they should do so only for good reason. In the event of litigation, the court may require them to explain why they decided not to adopt the recommended practice.

In addition, guidance notes are relevant to professional competence in that each surveyor should be up to date and should have informed himself of guidance notes within a reasonable time of their promulgation.

1 Introduction

1.1.1 The purpose of this Note is to provide practical guidance to RICS members on the skills needed and approach required when advising clients where there are failures by a landlord or tenant to comply with its legal obligations to repair, decorate, or reinstate alterations.

1.1.2 This Note deals primarily with commercial and industrial property in England and Wales and it is not intended to be a comprehensive guide to dilapidations. Where cross references are made in this document it is recommended that the surveyor makes himself familiar with the other documents to which reference is made.

1.1.3 Failure by a party to comply with a legal obligation can lead to a legal dispute. The need for a new edition of this Note arises from the adjustments which have been made to the Civil Procedure Rules in the civil courts of England and Wales (the CPR). Further references to the CPR will appear in this Note as appropriate.

1.1.4 The CPR encourages the parties to a dispute to exchange full information before proceedings are issued, to enable the parties to avoid litigation where possible and to support the efficient management of proceedings where litigation cannot be avoided. These objectives are addressed by way of pre-action protocols.

1.1.5 A draft of one such protocol (the Protocol) was composed by the Property Litigation Association (the PLA) and relates to a terminal schedule of dilapidations (see paragraph 2.2.1 below). A copy is annexed to this Note as Appendix 1. The PLA have consulted widely amongst interested parties, including the RICS, about the drafting of the Protocol and their views have been taken into account in the final draft annexed. Whether or not the Protocol is adopted by the Lord Chancellor's Department as an approved protocol, surveyors are encouraged to use it as a guide to good practice when addressing a terminal claim for dilapidations and to direct their client's attention to it.

1.1.6 For simplicity this Note assumes the tenant's failure to comply with its covenants. References to tenant can of course apply to the landlord in default. The parties are referred to in this Note as 'it' and the surveyor as 'he'.

2 Taking Instructions

2.1 General

2.1.1 Instructions relating to dilapidations should be taken in accordance with RICS Conduct and Disciplinary Procedures (the latest version of which came into force on 1 January 2003, and is available on the RICS website). Particular regard should be paid to paragraph 27.2.4, which requires notification of terms and conditions of engagement to be provided to the client in writing. Instructions in dilapidations matters are no different in this respect from any other instruction.

2.1.2 The surveyor also has an obligation to set out the basis of his fees in such a way that the client is aware of the financial commitment being made by instructing the surveyor.

2.1.3 The surveyor should bear in mind that he has duties to: the client; the RICS (in maintaining the reputation of the Institution and complying with its rules); and to any tribunal to which he may give evidence. As explained below, once a tribunal (typically, in dilapidations matters, the court) comes into play, the surveyor's primary responsibility is to that tribunal, rather than to the client or to the RICS. The obligations to the RICS and prospective obligations to the Tribunal should be brought to the attention of the client at an early stage, ideally with the preliminary instructions.

2.2 Types of Schedule

2.2.1 The surveyor may be instructed to prepare or consider a number of differing types of schedule. Instances include:

a) Detailing items of disrepair for service by the landlord during the currency of the lease (as a prelude to the landlord entering and carrying out the work and recovering the cost from the tenant if able to do so under the lease). During the course of the term, schedules may need to be prepared pursuant to provisions in the lease which entitle the landlord to enter the premises and undertake works should the tenant fail to carry them out (normally after a specified period).

Surveyors preparing such schedules should be careful to observe the exact provisions of the clause(s) in the lease. Otherwise, there is a risk of the tenant successfully asserting that the schedule is not served pursuant to the lease. The consequences of an inappropriately drafted schedule may be severe:

i) the landlord might be sued for trespass in the building;

ii) it may not prove possible for the landlord to recover all of the costs incurred in undertaking the works;

iii) the landlord could even be deemed to have peaceably re-entered the building, and forfeited the lease.

b) Identifying breaches of covenant for service with a notice under the Law of Property Act 1925, Section 146 (1) for the purposes of forfeiting the lease and/or obtaining damages. In this case, the surveyor should also be aware of the implications of the Leasehold Property (Repairs) Act 1938, Section 1(5). Both Acts are discussed further in paragraphs 5.9.3 and 5.9.4 below and the relevant sections are set out in Appendix 7.

c) A schedule prepared during the term on behalf of the tenant in order that it may comply with the terms of its lease.

d) A schedule served near to, at, or following expiry of the lease — referred to in this Note as a 'terminal schedule'.

2.2.2 For simplicity, the remainder of this Note is drafted in the context of a 'terminal' dilapidations claim scenario (see 2.2.1(d)). However, the principles identified are to some extent applicable to non-terminal situations.

2.3 The Landlord's Intentions

2.3.1 A landlord can only recover its monetary loss in a claim for damages. The intentions of the landlord with regard to the premises at or shortly after the end of the term are, therefore, of relevance to dilapidations issues. Dialogue between surveyors should take place with knowledge of such intentions — for example whether the building is to be demolished or physically altered in any way and if so, in what manner.

2.3.2 The extent to which advance dialogue between the parties is beneficial will be a matter of judgement. A certain amount of exchange of information can help towards resolving a dilapidations dispute. This argues for a dialogue between the parties prior to the expiry of the lease, or the formulation of a dilapidations issue.

2.4 Physical work vs damages

2.4.1 There may be advantages to both landlords and tenants in the repair works being carried out by the tenant prior to the end of the term. The landlord can more speedily market the premises, essential in a rapidly falling market. Equally, if the tenant chooses to do the work during the term:

- it will have control of the actual works and the timetable for those works;
- it may avoid a claim for loss of rent and interest; and
- it may be able to recover the VAT in circumstances in which, if the landlord carried out the works, it would have to pay a sum equivalent to irrecoverable VAT.

2.4.2 On the other hand, if the landlord has control of the works after the lease has expired, it will be able to dictate the works (within the meaning of the expired lease) and the timetable. The landlord can then recover the loss as damages. The tenant of course may be able to raise a defence to the claim for damages, for example pursuant to s18 (1) of the Landlord and Tenant Act, 1927 (see section 6.4), thereby paying less than if the works had been carried out during the term.

2.4.3 Whilst the financial and other more general circumstances of the parties, may dictate the strategy to be followed by each party, clients will often require advice from the surveyor on the appropriate course to follow.

2.5 Notices to Break

2.5.1 It is recommended that the surveyor takes care when asked to advise either a landlord or a tenant about a breach of covenant identified at the same time as a break clause is exercised pursuant to the lease. Some break clauses can only be exercised if the tenant has complied, or can comply by the expiry of the notice (dependent upon the wording of the lease), with all of its covenants contained within the lease. These are known as 'conditional break clauses'. In such circumstances, even a *de minimis* breach, for instance a minor building defect, can prevent the break clause being exercised or being effective. The surveyor is advised in all such cases to refer the matter to the client's solicitor.

3 Role of the Surveyor

3.1 General

3.1.1 There are four roles in which surveyors may be instructed in a dilapidations case: Expert Witness, Expert, Advisor and Negotiator. They often overlap. Considerable care will need to be taken in such cases, as discussed below.

3.1.2 Professional objectivity is required in all four roles. The surveyor should guard against exaggeration or understatement. Any subsequent litigation carries with it the danger of a heavy costs order against the party who exaggerates or understates, and this needs to be taken into account. Surveyors should be aware that pursuant to Part 44.3 of the CPR, the court has discretion as to whether costs are payable by one party to another:

'In deciding what order (if any) to make about costs, the court must have regard to all the circumstances, including [inter alia] the conduct of the parties.'

3.1.3 The rules go on to say that:

'The conduct of the parties includes:

a) conduct before, as well as during, the proceedings, and in particular the extent to which the parties followed any relevant pre-action protocol;

b) whether it was reasonable for a party to raise, pursue or contest a particular allegation or issue;

c) the manner in which a party has pursued or defended his case or a particular allegation or issue;

d) whether a claimant who has succeeded in his claim, in whole or in part, exaggerated the claim.'

3.2 Expert Witness

3.2.1 The role of an expert witness (whether appearing for one party or as a single joint expert) is dealt with in RICS Practice Statement and Guidance Note 'Surveyors Acting as Expert Witnesses' — second edition. The unusual circumstances in which departure from the Statement is possible are set out on page 4 of the Statement.

3.2.2 In the context of court proceedings, the surveyor's obligations are set out in Part 35 of the CPR and its accompanying Practice Direction. The surveyor will have to provide a Statement of Truth in any report prepared for the court as follows:

I confirm that insofar as the facts stated in my report are within my own knowledge I have made clear which they are and I believe them to be true, and that the opinions I have expressed represent my true and complete professional opinion.

3.2.3 Briefly stated, the obligations of an expert witness are as follows:

i) the overriding duty is to the court or tribunal before which he appears;

ii) he/she must give objective unbiased evidence. It follows that his evidence would be the same whether he acts for the tenant or the landlord, or appears as a single joint expert.

iii) he must ensure that instructions are confirmed in writing, and that the client understands a surveyor's obligations.

3.3 Expert

3.3.1 When a surveyor prepares a schedule of dilapidations, prices it, or prepares a valuation for the purpose of a dilapidations claim, he is doing so objectively, in a professional capacity. He is not preparing a tool for negotiation.

3.3.2 The schedule or valuation will appear in the landlord's claim, or in the tenant's defence in any subsequent legal proceedings. In the nature of things, it will then appear on the record. If it proves to be significantly different from the schedule or valuation finally presented to the court, the difference will have to be explained. An assertion that it was just a starting point is not likely to impress the court and may have an effect on the award of costs.

3.3.3 If a claim goes so far as to be the subject of a court judgment, the parties' conduct in the early stages of the claim will be examined. Surveyors should, therefore, conduct themselves within the spirit of the Protocol. Consequently, the production of a grossly exaggerated schedule of dilapidations or, conversely, an unrealistically understated assessment in response to a schedule is not acceptable. The same applies to valuation, pricing and the expression of any opinion. The manner in which a party then pursues or defends its position will also be taken into account. There may be cost consequences even against the successful party if its conduct is considered unreasonable.

3.3.4 An expert preparing the initial schedule or a valuation is not an expert witness. Whilst, therefore, the Practice Statement and the CPR will not bite until the court orders that expert witnesses may be called, nonetheless, from the outset, the surveyor should produce a schedule or valuation he considers he will later be able to present to the court in accordance with his obligations.

3.4 Advisor

3.4.1 The surveyor may be instructed simply to advise on strategy and tactics in a dilapidations claim. The surveyor should be influenced by the considerations relating to expert witnesses and to experts in advising the client.

3.5 Negotiator

3.5.1 Most claims do not end up in court. They are normally settled by negotiation. The negotiator will often be the surveyor who prepared the original schedule or valuation. In reaching a settlement the negotiator will consider the claim as a whole including time and costs in pursuing the claim. The surveyor should bear in mind that the total costs to both parties in relation to a claim taken through to trial will often exceed the value of the claim.

4 Information

4.1 The Lease

4.1.1 It is essential that the surveyor obtain a copy of the relevant lease in complete form with all plans and other attachments, coloured if originally coloured, together with all deeds of variation.

4.1.2 Additional information which may be necessary or desirable, depending on the circumstances, can include:

  i) scaled plans, coloured where appropriate;

  ii) licences or other consents for alterations, with plans and specifications;

  iii) any agreement for lease, if intended to survive the grant of the lease;

  iv) assignments and consents to assign;

  v) side letters or other written agreements;

  vi) schedules of condition, together with appropriate photographs;

  vii) schedules of fixtures and fittings;

  viii) the landlord's intentions for the property at, or shortly after, the termination of the tenancy (this may include details of works proposed to the premises);

  ix) any current planning consents or statutory notices relating to the property;

  x) any notices under the Landlord and Tenant Act 1954; and

  xi) any other indications of the landlord's intentions.

4.1.3 The surveyor needs to be satisfied that the documentation obtained is sufficient for the instructions to be discharged. Any questions as to authenticity may need to be addressed to the client or its legal advisor. All ambiguities in the documents or in instructions must be clarified as they arise.

4.2 Inspection

4.2.1 Whenever an inspection is to be undertaken before the lease expires, whether the tenant is in occupation or not, it is prudent to comply with the terms of the lease when making arrangements for access.

4.2.2 The surveyor is advised to acquaint himself with the RICS Guidance Notes 'Building Surveys and Inspections of Commercial and Industrial Property' and 'Surveying Safely' (the latter applying particularly when properties are empty).

4.2.3 It is advisable for the surveyor, at the time of inspection, to note the general standard of repair in the locality and whether similar properties are empty or boarded up. It is also advisable to note any changes to the nature of the area since the lease was granted.

4.2.4 Site notes, measurements or other transcriptions should be retained. When relevant, sketches with a north point should be made and photographs be taken. It is recommended that these be cross-referenced to the schedule and dated. If a video record is made, the same would apply.

4.2.5 The schedule must be capable of being interpreted by third parties. All parts of the property need to be clearly identified, together with alterations and improvements with cross-reference to licences where appropriate.

4.2.6 Further specialist input may be required from, for example, a services consultant or structural engineer. It is incumbent on the surveyor to ensure that the information provided by the specialist is capable of inclusion in the schedule.

4.2.7 A note may need to be made where further investigation or opening up is required. When this is justified, the agreement of the client and, where appropriate, the tenant should be obtained. It is incumbent on the surveyor to advise the client that if no defects are discovered then the cost of specialist inspections will not be recoverable.

5 The Schedule

5.1 Layout and Content

5.1.1 Before drafting a terminal schedule of dilapidations the surveyor should refer to the Protocol and Annex attached at Appendix 1 (see particularly paragraph 3.4 as to separation of various breaches of covenant, and the preferred layout of a schedule of dilapidations in the Annex). An example of a schedule of dilapidations is also attached, see Appendix 3.

5.2 Scott Schedule vs Schedule of Dilapidations

5.2.1 The schedule of dilapidations attached to the Protocol does not provide any space for the comments of the tenant's surveyor. It is highly desirable nonetheless that any such comments should be made against each item, perhaps even when the tenant's surveyor makes notes on site. To facilitate this, the landlord's surveyor should provide the Scott Schedule form of the schedule of dilapidations (see section 5.8 below) to the tenant's surveyor as well as the form which will have been served on the tenant itself, so that the tenant's surveyor is enabled, and encouraged, to make comments on the same document.

5.2.2 In some claims this may seem to be unduly heavy-handed. Nonetheless, with the increasing emphasis on the claim being complete to begin with, it will avoid a great deal of transposition work later in the day if it subsequently appears that the matter is moving towards trial.

5.3 Repair

5.3.1 Repairing covenants may either be express, or may be implied, for example 'to use the premises in a tenant-like manner', and 'not to commit waste'.

5.3.2 An account of the basic principles governing repair is beyond the scope of this Note. Reference to the various texts listed in the Bibliography (Appendix 6) is recommended.

5.4 Decoration

5.4.1 Leases normally provide for the redecoration of premises within specified cycles throughout the term (typically three years for external parts and five or seven years for internal parts) and in the final six months or final year of the term.

5.4.2 The provisions of these covenants should be carefully studied as they do vary. The number of coats of paint may be specified. Reference may be made to surfaces which are 'previously', 'usually' or 'ought to be' decorated. There may also be additional obligations to 'treat', 'polish' or 'restore' certain surfaces.

5.5 Reinstatement

5.5.1 It may be evident to the surveyor, either from inspection of the premises or from the documentation, that the premises have been altered. In such circumstances enquiry may need to be made of the client and/or it's legal advisor to what extent the tenant is obliged to reinstate.

5.5.2 The surveyor, in seeking to identify alterations, will have regard to:
   i) obvious differences in construction and materials;
   ii) materials which are inconsistent with the age of the building;
   iii) parts of the property which directly identify with the trade or occupation of the tenant (for example an extension constructed to store chemicals);
   iv) plans, photographs or other documentary evidence; and
   v) the existence of partitions and fitting out.

5.6 Statutory Obligations

5.6.1 Leases normally include covenants requiring the tenant to comply with and carry out works required by the provisions of any relevant statute or regulation.

5.6.2 Many statutory obligations arise only in respect of occupied premises (e.g. fire regulations). For this reason breaches of statutory obligations are best addressed by the landlord during the term.

5.7 Costing

5.7.1 The schedule of dilapidations will need to be costed if it is anticipated that the appropriate remedy is by way of a payment in damages.

5.7.2 A schedule of dilapidations should be priced with due reference to reliable and appropriate cost information which is available from a number of sources, for example: (i) current Building Cost Information Service data and other recognized price books (to which the appropriate regional variations should be applied); (ii) relevant and recent tender price information (on projects of a similar nature and size and envisaged by the claim); and (iii) the result of a competitive tender exercise (which could be conducted on the basis of a full specification of works derived from the schedule of dilapidations).

5.7.3 For larger and more complicated claims it may be advisable for surveyors to recommend to the client that the services of a quantity surveyor be engaged to complete the pricing process, the cost of which would normally be recoverable as part of the cost of preparing the schedule.

5.8 Format

5.8.1 It is vital that the format of a schedule of dilapidations is such that it can be edited. Nowadays, this usually means a computerised format, preferably one which is commonplace, readily capable of being edited, and which carries out calculations automatically.

5.8.2 During dialogue on a schedule of dilapidations between the parties (see Section 7.0), the schedule may require extensive editing. For example, a Scott Schedule is essentially an expanded version of the original schedule of dilapidations. Both parties' comments and costings will appear on the Scott Schedule. Inevitably, in the process of dialogue, some of these comments and costings will change. One surveyor may convince the other that a point he has relied on is misconceived, or that a costing is exaggerated or understated.

5.8.3 A systematic method of maintaining versions of the document, and exchanging edits is recommended. This may well be done electronically. One surveyor is charged with maintaining versions of the document. These should each be clearly marked with:

i) the nature of the document;

ii) whether, for example, it is a schedule of dilapidations or a Scott Schedule;

iii) the version number;

iv) the revision number;

v) the person who maintains the master copy;

vi) the person by whom it has been revised; and

vii) the date upon which the revision took place.

5.8.4 These factors should appear on the head sheet of the document, in something like the following format:

Document	Scott Schedule
Version	A
Revision	3
Master maintained by	Landlord's surveyor
Latest edits by	Tenant's surveyor
Date of latest edits	31 March 2002

5.8.5 If done electronically this approach suggests the creation of a specific folder on the landlord's surveyor's computer with the name of the property concerned and perhaps with subfolders for Schedules of Dilapidations and for Scott Schedules. These should then be named Revision 1, Revision 2 etc, or given some other systematic nomenclature to record the history of the final Scott Schedule.

5.8.6 Obviously, the manual equivalent of this process is perfectly acceptable as a matter of law and professional practice. One approach to doing the same task by non-electronic methods is to use the lawyers' 'travelling draft' approach. In other words, the first version of the Scott Schedule is typed. Then edits by the next surveyor to consider it are made by hand in red. The next set is done in blue, the next set in green and so on. At some stage during the process, a fair copy of those alterations is produced.

5.8.7 The electronic method of working suggests that revision mode should be used — if it is, edits can be readily identified on the computer document — not only where and what they are, but who made the alteration. Where needed for clarification, comments could be attached to each edited cell to indicate why the alteration is being made. If a process of this sort is agreed between surveyors, it could enable the dialogue to proceed rapidly, having regard to the tight time limits in the protocol. If the dialogue is conducted electronically, exchange of documents by e-mail (as opposed to floppy discs or paper copies) will speed the dialogue.

5.9 Service

5.9.1 The schedule of dilapidations will usually be served by the client's solicitor. This is because formalities may be required to be followed and observed either pursuant to the lease or statute.

5.9.2 For instance, the lease may stipulate that any notices served under the lease must be served by recorded delivery or be served by hand.

5.9.3 If the client wishes to reserve the right to forfeit the lease upon the tenant's failure to carry out the required works, the schedule must be annexed to a notice served under s146 of the Law of Property Act, 1925 (known as a 's146 notice'), see Appendix 7. A right to re-enter or forfeit property is not enforceable unless the s146 notice complies with the requirements contained within this section of the Act.

5.9.4 Further, if the lease is for a term of seven years or more, of which three years or more remain unexpired, then a landlord cannot claim damages for a breach of a repairing covenant under the lease unless a s146 notice is served (annexing the schedule to it) containing a statement to the effect that the tenant is entitled to serve a counter-notice claiming the benefit of the Leasehold Property (Repairs) Act 1938 (s1 of that Act) — see Appendix 7. If a counter-notice is served by the tenant within 28 days of service of the s146 notice, no proceedings can be commenced against the tenant in relation to the breach of covenant without the formal leave of the court. In obtaining that leave, the landlord will be required to prove one of the five grounds in s1(5) of the 1938 Act.

6 The Claim

6.1 Introduction

6.1.1 The claim should be quantified in a separate document indicating clearly how it is made up (see Section 4 of the Protocol).

6.1.2 The surveyor's original schedule will obviously be one of the documents appended to the claim. It is normally also the surveyor's responsibility to prepare, where appropriate, a summary of the figures which make up the claim, and of which the total of the schedule is only a part.

6.1.3 The claim summary ought to be set out on one sheet which, like the schedule itself, is expandable, so that the other party's surveyor can provide corresponding figures in due course. The summary could be a worksheet in a computer workbook, which draws in the cost of the various headings of work from other worksheets.

6.2 Consequential costs

6.2.1 The following items will often be added, if appropriate, to the schedule figures to make up the total claim:

   i) fees for the preparation of the schedule;

   ii) fees for the supervision of the work;

   iii) VAT;

   iv) loss of rent and rates;

   v) loss of service charges and/or insurance premium;

   vi) fees in relation to negotiation of the schedule; and

   vii) solicitor's costs and interest (subject to consultation with the client's solicitor).

6.3 VAT

6.3.1 VAT is a subject beyond the scope of this Note, but it is obvious that if a landlord can recover the VAT element in the claim from HM Customs and Excise then it cannot also claim the equivalent from the outgoing tenant. If the landlord cannot claim back VAT, then an equivalent sum will normally be a legitimate part of the claim. (See Appendix 6, Bibliography for details of an article on this subject).

6.4 Valuation

6.4.1 The valuation called for in a dilapidations claim is often loosely referred to as 'a s18 valuation'. This is because s18 (1) of the Landlord and Tenant Act 1927 (see Appendix 7) provides a statutory cap on damages for disrepair. A landlord cannot recover for breach of a covenant to repair beyond the loss to its reversionary interest. As regards other heads of claim the common law also provides that a landlord cannot recover more than it has lost. In dilapidations this is also expressed as the diminution in the value of the landlord's reversion. It is, therefore, better to refer to the valuation required as a 'diminution valuation'.

6.4.2 A judgment will have to be made at the claim stage as to whether a diminution valuation should be prepared and appended to the claim. In many simple claims, it will be of no significance. The claim is simply the cost of the work.

6.4.3 At the other extreme, it may be obvious that a diminution valuation is significant. For example, where part of the premises is to be entirely changed, regardless of whether the tenant has complied with its covenants or not, then it is obvious that the diminution principle will have an impact. In such a case a diminution valuation should be appended to the claim in accordance with the requirements of the CPR.

6.4.4 In more marginal cases, it might be said that it is for the tenant to plead the diminution point as a defence, and therefore, that it is for the tenant, if it considers that diminution has an impact, to prepare a diminution valuation as part of the defence to the claim.

6.5 Supersession

6.5.1 'Supersession' is the term used where the need for an item of work is overtaken by the probable occurrence of another.

6.5.2 The surveyor assessing diminution in the value of the landlord's reversion needs to distinguish between items that will and will not be superseded. For example, if a wall needs to be re-plastered in fact and pursuant to the lease, then this item will legitimately appear in the schedule of dilapidations and be costed. The tenant's surveyor may agree that it falls within the covenant to repair, and he may agree with the costing. However, he may say that because, for instance, the wall is due to be demolished or altered in some way by the landlord, the work of replastering is superseded by the intention to alter or demolish the wall.

6.5.3 It is for the surveyor to determine which items are superseded. These should be recorded. One way of doing this is by the addition of further columns to the Scott Schedule.

6.5.4 The Scott Schedule and the supersession schedule, taken together, will indicate the entire nature of the dispute on an item by item basis, whether the dispute is about the aptness of an item, its costing or whether it is to be superseded. This will assist the Judge (not to mention the parties in the run up to trial) in assessing the strengths and weaknesses of the parties in relation to each item.

6.5.5 The supersession schedule does not avoid the need for a diminution valuation, but is part of the thinking underlying that valuation.

7 Dialogue

7.1 Timetable

7.1.1 The Protocol contains recommended timing for the tenant's response to the claim letter in relation to a terminal schedule of dilapidations, and the subsequent progress of negotiations. At all times, the surveyor is advised to seek to advance the dialogue, through correspondence and meetings, in a timely manner. This applies whether the claim is in respect of a terminal schedule or otherwise.

7.2 The Scott Schedule

7.2.1 The suggested structure of a Scott Schedule is set out in Appendix 4 to this Note. A partial sample of a Scott Schedule is given in Appendix 5. The objective of the Scott Schedule is to set out for the parties (and, if necessary, the court) the nature and extent of the dispute on the contractual claim, on an item-by-item basis. Paragraph 5.8.2 mentions that a Scott Schedule is essentially an expanded version of the original schedule of dilapidations. The original schedule will have a series of items on it, often considerable in number. The Scott Schedule must address the parties' positions on each of these items individually. The first four columns — Item number; Clause number; Breach complained of; and Remedial works required — will all have appeared on the original schedule of dilapidations, as will column 7 — 'Landlord's costing'. The additional columns introduced at the Scott Schedule stage are numbered 5, 6, 8, 9 and 10: Tenant's comments on the landlord's item; Landlord's comments on the response; Tenant's costing of the landlord's item; Landlord's costing of the tenant's item (which may be different from the landlord's); and Tenant's costing of its own item.

7.2.2 The thrust of the four cost columns — numbers 7 to 10 — is to identify the character of the dispute on each item. After all, there are four possible positions that a tenant might take in relation to a landlord's item:

1) agreement;

2) agreement with the item, but not with the costing;

3) agreement with the costing but not with the item; and

4) agreement with neither the costing nor the item.

7.2.3 The four columns of figures are intended to clarify into which category the dispute on each item falls. If the matter reaches trial, the figures provided under 2 and 3 above will be particularly helpful to the judge. If the item is agreed, he can simply determine the costing. If the costing is agreed, he can just decide whether the item is appropriate.

7.2.4 It is essential that there should be a reference number for each item. As the dialogue over the Scott Schedule progresses, a number of items may be eliminated from the claim. They should nonetheless continue to appear on the Scott Schedule, no doubt with zeros against them, in order to record the history of the negotiations between the parties and the reason why items have been eliminated.

7.3 Responses and meetings

7.3.1 Both the court and the RICS Expert Witness Practice Statement and Guidance Note 'Surveyors Acting as Expert Witnesses' encourage dialogue between parties to a dispute. In order to achieve this, surveyors of like disciplines should meet during the course of the dispute, in order to clarify the nature of the dispute and if possible, to settle aspects of it. This process will often reveal that the areas of disagreement are relatively few, and the areas of agreement many. This often permits the claim to be settled or, if it is not settled, to ensure that the dispute focuses on those matters that are truly not agreed, as opposed to those areas where the parties do not understand one another.

7.3.2 Due to the very high costs involved, it must be the parties' objective, as well as the court's, that the matter should be settled instead of being tried, if at all possible.

7.4 Alternative Dispute Resolution

7.4.1 Most disputes involving dilapidations are settled between the parties out of court. However, a few do lead to a court case. Pursuant to the CPR, the court and not the parties will actively manage the case listed for hearing and in so doing will encourage the parties to use an Alternative Dispute Resolution procedure (ADR) (Part 1.4(2) (e) of the CPR). ADR is described in the glossary to the court rules as the 'collective description of methods of resolving disputes otherwise than through the normal trial process'. The surveyor may obtain further information about ADR through the RICS Dispute Resolution Faculty.

Appendix 1

THE PROTOCOL AND ANNEX

Property Litigation Association

Protocol for Terminal Dilapidations Claims for Damages

1 Introduction

1.1 This protocol has been drafted by the Law Reform Sub-Committee (the "Committee") of the Property Litigation Association. It is hoped that it will be adopted by the Lord Chancellor's Department and become a protocol under the Civil Procedure Rules (the "CPR"), the rules by which civil litigation in England and Wales is conducted. This protocol relates to claims for damages for breach by tenants of repair obligations in leases at the expiry of the lease. It sets out the Committee's views on effective and appropriate standards to be adopted for the efficient conduct of pre-action litigation.

1.2 It is not the purpose of this protocol to define "dilapidations". The work that may be required will depend on the definition adopted by the parties in each particular case. However, as a guide, "dilapidations" might be said to be a reference to a state of disrepair in a property, or a condition of that property that requires work to rectify it, and there is a legal liability to remedy, or undertake, that work.

1.3 This protocol is not intended to be an exhaustive or mandatory list of steps or procedures to be followed. Those will be determined by the facts of each case. In deciding the exact steps and procedures to be adopted regard should also be had to the Overriding Objective as set out in CPR Part 1 and the Practice Direction - Protocols.

1.4 This protocol is intended to improve the pre-action communication between landlord and tenant by establishing a timetable for the exchange of information relevant to the dilapidations dispute and by setting standards for the content of claims and correspondence and the conduct of pre-action negotiations.

1.5 Compliance with the protocol should enable both landlords and tenants to make an informed judgment on the merits of their cases as soon as possible. The aim is to increase the number of pre-action settlements. If proceedings are commenced, then notwithstanding that the protocol has not yet been "approved" under the CPR, the court may be invited by any party to treat the standards set out in this protocol as the normal reasonable approach to pre-action conduct when the court considers issues of costs and other sanctions under the CPR. When doing so, the court should be concerned with substantial compliance and not minor departures, e.g. failure by a short period to provide relevant information. In addition, minor departures should not exempt the "innocent" party from following the protocol. The court may also be invited to consider the effect of non-compliance on the other party when deciding whether to impose sanctions.

2002 © Copyright: Property Litigation Association

2 Overview of Protocol — General Aim

The objectives of this protocol are:

a) to encourage the exchange of early and full information about the prospective legal claim;

b) to enable parties to avoid litigation by agreeing a settlement of the claim before the commencement of proceedings;

c) to support the efficient management of proceedings where litigation cannot be avoided.

The Protocol

3 The Schedule

3.1 Generally, the landlord shall serve a schedule. It shall indicate the breaches of the tenant's repairing obligations which have not been cured at the determination of the tenancy and state what in the opinion of the landlord or the surveyor is necessary to put the premises into repair in accordance with the terms of the lease. If the landlord has carried out the work necessitated by the tenant's breach, a schedule may not be appropriate.

3.2 The schedule shall be served within a reasonable time. A "reasonable time" will vary from case to case but generally will be not more than 2 months after the determination of the tenancy.

3.3 The landlord may serve a schedule before the determination of the tenancy.

3.4 If it is intended to claim for breaches of reinstatement or other obligations, these should be listed separately (but in one document, if appropriate) and should (where appropriate) identify any notices served by the landlord requiring reinstatement works to be undertaken.

3.5 Generally the schedule should be in the form of the Annex attached.

3.6 If possible the schedule should also be provided by way of computer disk or similar form to enable the tenant's comments to be incorporated in the one document.

4 The Claim

4.1 The claim should be quantified in a separate document (see paragraph 4.4 below) indicating clearly how that claim is made up.

4.1.1 Generally, if the claim is based on the cost of works, it should be fully quantified and substantiated. For example, each item of expenditure should, where possible and/or relevant, be supported by either an invoice or estimate.

2002 © Copyright: Property Litigation Association

4.1.2 If the landlord has carried out the work, it is not required to provide a valuation under Section 18(1) of the Landlord and Tenant Act 1927 ("section 18(1)"). If the landlord has not carried out the work but intends to, it must state when it intends to do the work, and what steps it has taken towards getting the work done, e.g. preparing a specification or bills of quantities or inviting tenders and the landlord should provide a section 18(1) valuation unless, in all the circumstances, it would be reasonable not to. If the landlord does not intend to carry out the work, then it must provide a section 18(1) valuation which should provide sufficient costing of the works to demonstrate that the cost of the works would exceed the section 18(1) valuation. The nature and detail of the section 18(1) valuation will depend upon the circumstances in each case.

4.2 All aspects of the claim including the VAT status of the landlord, if appropriate, should be set out.

4.3 If the claim includes any other losses such as, (a) projected surveyor's fees for negotiating the claim, (b) for lost profits, (c) preliminaries, (d) overheads and loss of rent/service charges, and (e) surveyor's fees in preparing the schedule, these must be set out in detail, fully quantified and substantiated (whether in the schedule or in the claim letter). If section 18(1) applies to these types of claim, then they must be treated in the same way as in paragraphs 4.1.2 and 5.3.

4.4 The claim letter must be sent and should generally contain the following information:

- the landlord's full name and address;
- the tenant's full name and address;
- a clear summary of the facts on which the claim is based;
- the schedule referred to above;
- a clear summary of the claim which may include the cost of the works, the consequential costs and fees, VAT, loss of rent and other losses (including any sums paid to a superior landlord);
- any documents relied upon or required by this protocol, including copies of any receipted invoices or other evidence of such costs;
- confirmation that the landlord and/or its professional advisers will attend a meeting or meetings as proposed under section 7 below;
- a date (being a reasonable time) by which the tenant should respond.

5 The Response

5.1 The tenant must respond to the claim letter in a reasonable time. In the usual case 2 months should be adopted as a reasonable time.

5.2 The tenant should respond having regard to the schedule provided by the landlord, where appropriate, in sufficient detail to enable the landlord to understand clearly the tenant's views on each item of claim.

2002 © Copyright: Property Litigation Association

5.3 If the tenant relies upon section 18(1), it should state its case for so doing and provide a valuation where appropriate. If appropriate, and if not provided by the landlord, the tenant should also request the landlord to provide proportionate and reasonable disclosure of documents relevant to the landlord's intention to carry out works to the premises.

6 Disclosure of Documents

Disclosure will generally be limited to the documents required to be enclosed with the claim letter and the tenant's response. The parties can agree that further disclosure may be given. If either or both of the parties consider that further disclosure should be given but there is disagreement about some aspect of that process, they may, through their solicitors, be able to make an application for pre-action disclosure under CPR Part 31.

7 Negotiations

7.1 The landlord and tenant and/or their respective professional advisers are encouraged to meet before the tenant is required to respond to the claim letter and must generally meet within 1 month of service of the tenant's response. The meetings will be without prejudice and preferably on site, to review the schedule to ensure that the tenant fully understands all aspects of the landlord's claim. The parties shall seek to agree as many of the items in dispute as possible.

7.2 In a complex matter it may be necessary for more than one site visit or without prejudice meeting between the parties to take place. These ought to be conducted without unnecessary delay.

8 Stocktake

Where a claim is not resolved when the protocol has been followed, the parties might wish to carry out a "stocktake" of the issues in dispute, and the evidence that the court is likely to need to decide those issues, before proceedings are started.

9. Alternative Dispute Resolution

Both parties are to explore the possibilities of mediation or other alternative dispute resolution process.

July 2002
Law Reform Sub-Committee
Property Litigation Association

2002 © Copyright: Property Litigation Association

Annex
Schedule of Dilapidations
[The Property]

This schedule has been prepared by [name, individual and firm], upon the instructions of [name the landlord]. It was prepared following [name i.e. same name as above]'s inspection of the premises known as [property] on [date].

It records the works required to be done to the premises in order that they are put into the condition the premises should have been put if the tenant [name] had complied with its covenants contained within its lease of the premises dated [date].

The covenants of the said lease with which the tenant should have complied are as follows:

[Set out clause number of the lease and quote the clause verbatim].

The following schedule contains:

- reference to the specific clause (quoted above) under which the repairing obligation arises,
- the breach complained of,
- the remedial works suggested by the landlord's surveyor [name i.e. same name as above] as suitable for remedying the breach complained of,
- the landlord's view on the cost of the works.

The schedule contains the true views of [name, i.e. the same name as above] being the surveyor appointed/employed by the landlord to prepare the schedule.

Upon receipt of this schedule the tenant should respond as required by the Protocol to enable the landlord to understand clearly the tenant's views on each item of claim.

2002 © Copyright: Property Litigation Association

1 Item No.	2 Clause No.	3 Breach complained of	4 Remedial works required	5 Landlord's costing

DATED [............................]

SIGNED [............................]
[Name and address of surveyor appointed by landlord]

2002 © Copyright: Property Litigation Association

Appendix 2

SCHEDULE OF OTHER REFERENCES

Professional Conduct: 'Rules of Conduct and Disciplinary Procedures' (in force 1 January 2003)

RICS Practice Statement and Guidance Note: 'Surveyors Acting as Expert Witnesses', 2nd edition

RICS Guidance Note: 'Building surveys and Inspections of Commercial and Industrial Property'

RICS Guidance Note: 'Surveying safely: a commitment to personal safety'

Appendix 3

EXAMPLE OF A SCHEDULE OF DILAPIDATIONS

Case: STREAMSIDE, ROBIN HOOD WAY, NOTTINGHAM
REPAIR AND DECORATION

Section No 1: page 1 of 2

1	2	3	4	7
Item no	Clause no	Breach complained of	Remedial works required	Landlord's costing
Car parking area				
1	4(6)(a)	Floodlighting is provided within the planting area to light the front elevation of the building. Floodlight fittings are corroded and broken. None of the 4 No. fittings are in working order.	Renew 4 No. floodlight fittings and check switchgear and wiring.	2,000.00
2	4(6)(a)	Car parking and hardstanding to east of the north wing. Precast concrete paving slabs to paths adjacent to building. 30 No. slabs are cracked and broken. 50 No. slabs are displaced and uneven.	Take up and dispose of damaged slabs. Lay new slabs to match existing. Take up and re-lay uneven and displaced slabs.	1,600.00
3	4(6)(a)	Grass covered banks between perimeter paving and footpath to roadway. Grass has died to 25m².	Prepare top soil base and lay 25m² of new turf.	450.00
4	4(6)(a)	Gas access chamber adjacent to directors' parking bays. Chamber is damaged.	Repair damaged chamber.	600.00
5	4(6)(a)	Tarmacadam surfacing to car park areas with thermoplastic markings to car park bays. There is localized damage to tarmac surface including pot holes approx 300mm diameter particularly adjacent to gullies.	Patch repair tarmac to damaged areas (allow 15m²).	875.00
6	4(6)(a)	The markings to car parking bays are worn and missing in places.	Burn off all residual markings and apply new thermoplastic markings to all parking bays; re-design and set out to standard size parking bays.	3,500.00
Internal: ground floor – west wing				
7	4(6)(a)	4 No. new infill bays following removal of tenant blockwork. Glazing not tinted to match existing.	Replace glass with tinted glass – 16 No. windows in total.	3,200.00
8	4(7)	Versatemp units installed by tenant.	Refix units that have not been securely mounted on external walls.	1,500.00
9	4(7)	Computer trunking skirting to perimeter. Not laid against the floor leaving uneven gap.	Take up. Refit at lower level to close up the gap; make good finishes; redecorate.	2,750.00

Section No 1: page 2 of 2

1	2	3	4	7
Item no	Clause no	Breach complained of	Remedial works required	Landlord's costing
10	4(7)	Ceiling tiles: damp staining already showing from roof leaks above. 2 No. tiles missing.	Replace missing and damaged tiles.	100.00
11	4(7)	Smokers room window frames heavily stained from nicotine etc.	Replace cover frames left by tenant when this area was formerly a smoking room.	950.00
12	4(7)	Plasterwork is damp damaged particularly at high level in lobbies adjacent to curtain wall enclosure.	Hack off damp damaged plasterwork and make good. Repair wall surfaces and prepare for redecoration.	3,685.00
13	4(7)	Decorative finishes to wall surfaces have been damaged and are soiled.	Apply mist coat and two full coats emulsion paint finish to all previously decorated surfaces.	2,025.00
		Escape stairs to west wing		
14	4(6)(b)	Plasterwork is damp damaged particularly at high level in lobbies adjacent to curtain wall enclosure.	Hack off damp damaged and cracked plasterwork and make good.	3,685.00
15	4(6)(b)	There is extensive cracking to plastered surfaces.	Hack off damp damaged cracked plasterwork and make good.	Inc above
16	4(6)(b)	Decorative finishes to wall surfaces have been damaged and are soiled.	Repair wall surfaces and prepare for redecoration. Apply mist coat and two full coats emulsion paint finish to all previously decorated surfaces.	2,025.00
17	4(6)(b)	Ceiling tiles are generally discoloured. Many tiles are damaged, chipped or stained	Renew all suspended ceiling tiles (replacement of damaged tiles only will result in a patchy and inconsistent appearance).	3,785.00
18	4(6)(b)	Carpets have widespread staining as a result of spillages and poor maintenance.	Take up and dispose of all carpet tiles. Supply and lay new carpet tiles of a grade and quality to match existing tiles.	5,600.00
TOTALS – **REPAIR AND DECORATION**				38,330.00

15/10/02   Appendix 3

Appendix 4

RECOMMENDED FORM OF SCOTT SCHEDULE

Scott Schedule. Case: PROPERTY ADDRESS
REPLACE THIS TEXT WITH THE CATEGORY OF WORK

Section No 1: page 1 of 2

1	2	3	4	5	6	7	8	9	10
						Landlord's item		Tenant's item	
Item no	Clause no	Breach complained of	Remedial works required	Tenant's comments	Landlord's comments	Landlord's costing	Tenant's costing	Landlord's costing	Tenant's costing
		Replace this text with the sub-category of work described in this section of the sheet							
1		Replace this text with the nature and location of the defect.	Replace this text with the works required to remedy the defect.			0.00	0.00	0.00	0.00
2		Replace this text with the nature and location of the defect.	Replace this text with the works required to remedy the defect.			0.00	0.00	0.00	0.00
3		Replace this text with the nature and location of the defect.	Replace this text with the works required to remedy the defect.			0.00	0.00	0.00	0.00
4		Replace this text with the nature and location of the defect.	Replace this text with the works required to remedy the defect.			0.00	0.00	0.00	0.00
5		Replace this text with the nature and location of the defect.	Replace this text with the works required to remedy the defect.			0.00	0.00	0.00	0.00
6		Replace this text with the nature and location of the defect.	Replace this text with the works required to remedy the defect.			0.00	0.00	0.00	0.00
		Replace this text with the sub-category of work described in this section of the sheet							
7		Replace this text with the nature and location of the defect.	Replace this text with the works required to remedy the defect.			0.00	0.00	0.00	0.00
8		Replace this text with the nature and location of the defect.	Replace this text with the works required to remedy the defect.			0.00	0.00	0.00	0.00
9		Replace this text with the nature and location of the defect.	Replace this text with the works required to remedy the defect.			0.00	0.00	0.00	0.00
10		Replace this text with the nature and location of the defect.	Replace this text with the works required to remedy the defect.			0.00	0.00	0.00	0.00

1	2	3	4	5	6	7	8	9	10
						Landlord's item		Tenant's item	
Item no	Clause no	Breach complained of	Remedial works required	Tenant's comments	Landlord's comments	Landlord's costing	Tenant's costing	Landlord's costing	Tenant's costing
11		Replace this text with the nature and location of the defect.	Replace this text with the works required to remedy the defect.			0.00	0.00	0.00	0.00
12		Replace this text with the nature and location of the defect.	Replace this text with the works required to remedy the defect.			0.00	0.00	0.00	0.00
13		Replace this text with the nature and location of the defect.	Replace this text with the works required to remedy the defect.			0.00	0.00	0.00	0.00
			Replace this text with the sub-category of work described in this section of the sheet						
14		Replace this text with the nature and location of the defect.	Replace this text with the works required to remedy the defect.			0.00	0.00	0.00	0.00
15		Replace this text with the nature and location of the defect.	Replace this text with the works required to remedy the defect.			0.00	0.00	0.00	0.00
16		Replace this text with the nature and location of the defect.	Replace this text with the works required to remedy the defect.			0.00	0.00	0.00	0.00
17		Replace this text with the nature and location of the defect.	Replace this text with the works required to remedy the defect.			0.00	0.00	0.00	0.00
18		Replace this text with the nature and location of the defect.	Replace this text with the works required to remedy the defect.			0.00	0.00	0.00	0.00
19		Replace this text with the nature and location of the defect.	Replace this text with the works required to remedy the defect.			0.00	0.00	0.00	0.00
20		Replace this text with the nature and location of the defect.	Replace this text with the works required to remedy the defect.			0.00	0.00	0.00	0.00
21		Replace this text with the nature and location of the defect. If you need more lines than this, use the fill handle to extend the range downwards.	Replace this text with the works required to remedy the defect.			0.00	0.00	0.00	0.00
TOTALS – CATEGORY NAME *[MAKE SURE THE FORMULA REFERENCES THE RIGHT RANGE OF CELLS]*						0.00	0.00	0.00	0.00

Appendix 5

EXAMPLE OF SCOTT SCHEDULE

Scott Schedule. Case: STREAMSIDE, ROBIN HOOD WAY, NOTTINGHAM
REPAIR AND DECORATION

Section No 1: page 1 of 4

1	2	3	4	5	6	7	8	9	10
						Landlord's item		Tenant's item	
Item no	Clause no	Breach complained of	Remedial works required	Tenant's comments	Landlord's comments	Landlord's costing	Tenant's costing	Landlord's costing	Tenant's costing
				Car parking area					
1	4(6)(a)	Floodlighting is provided within the planting area to light the front elevation of the building. Floodlight fittings are corroded and broken. None of the 4 No. fittings are in working order.	Renew 4 No. floodlight fittings and check switch gear and wiring.	Item agreed – allow for retubing 4 No. floor light fittings and check electrical wiring to ensure safe operation. Landlord's claim is excessive.	Renewal is closer to the standard contemplated by the covenant.	2,000.00	1,600.00	1,600.00	1,600.00
2	4(6)(a)	Car parking and hardstanding to east of the north wing. Precast concrete paving slabs to paths adjacent to building. 30 No. slabs are cracked and broken. 50 No. slabs are displaced and uneven.	Take up and dispose of damaged slabs. Lay new slabs to match existing. Take up and re-lay uneven and displaced slabs.	Item agreed – numerous broken and displaced precast concrete paving flags are noted to the access path to the north wing east elevation. Landlord's claim excessive.	£1,200 will not meet the requisite standard.	1,600.00	1,200.00	1,600.00	1,200.00
3	4(6)(a)	Grass covered banks between perimeter paving and footpath to roadway. Grass has died to 25m².	Prepare top soil base and lay 25m² of new turf.	At the time of the lease term expiry an area of approximately 25m² of the grassed area was noted to be bare and/or missing. The landlord has failed to maintain the condition of the soft landscaping since lease expiry in January 2002. The soft landscaping is now heavily weed growth affected. Landlord's claim is excessive. The tenants should pay only a contribution towards the reinstatement of the landscaping back to the condition as witnessed at lease term expiry.	The grassed area should have been put into and kept in the condition contemplated by the covenant.	450.00	450.00	300.00	150.00

10/09/02

Appendix 5

Section No 1: page 2 of 4

1	2	3	4	5	6	7	8	9	10
						Landlord's item		Tenant's item	
Item no	Clause no	Breach complained of	Remedial works required	Tenant's comments	Landlord's comments	Landlord's costing	Tenant's costing	Landlord's costing	Tenant's costing
4	4(6)(a)	Gas access chamber adjacent to directors' parking bays. Chamber is damaged.	Repair damaged chamber.	A small gas chamber is located within the paving flag access path adjacent to the directors' car park. Extremely minor impact damage is noted to the concrete haunching/bedding to the gas chamber housing. Landlord's claim is excessive.	The minor damage nonetheless requires a major repair.	600.00	100.00	600.00	100.00
5	4(6)(a)	Tarmacadam surfacing to car park areas with thermoplastic markings to car park bays. There is localized damage to tarmac surface including pot holes approx 300 mm diameter particularly adjacent to gullies.	Patch repair tarmac to damaged areas (allow 15m²).	Item disagreed – generally the tarmacadam hardstanding is in reasonable order with the exception of small number of pot holes. The areas identified in the landlord's schedule (15m²) are excessive. Approximately 10 No. pot holes to all car parks require attention in order to return them into good condition. Landlord's claim is excessive. Allow for making good 10 No. pot holes approx 500m² by 150mm deep.	Agreed, and cost reduction agreed.	875.00	875.00	450.00	450.00
6	4(6)(a)	The markings to car parking bays are worn and missing in places.	Burn off all residual markings and apply new thermoplastic markings to all parking bays; re-design and set out to standard size parking bays.	Item agreed – the road markings to the rear car park area are aged, weathered and require to be renewed. The landlord's claim is excessive.	£2,500 will not meet the requisite demand.	3,500.00	2,500.00	3,500.00	2,500.00
				Internal: ground floor – west wing					
7	4(6)(a)	4 No. new infill bays following removal of tenant blockwork. Glazing not tinted to match existing.	Replace glass with tinted glass – 16 No. windows in total.	Item agreed – the glazing panes to the new replacement curtain walling to the ground floor west wing do not match the tint and colour of the original curtain walling sections.	Agreed.	3,200.00	3,200.00	3,200.00	3,200.00
8	4(7)	Versatemp units installed by tenant.	Refix units that have not been securely mounted on external walls.	Item agreed.	Agreed.	1,500.00	1,500.00	1,500.00	1,500.00

10/09/02

Appendix 5

Section No 1: page 3 of 4

1	2	3	4	5	6	7	8	9	10
						Landlord's item		Tenant's item	
Item no	Clause no	Breach complained of	Remedial works required	Tenant's comments	Landlord's comments	Landlord's costing	Tenant's costing	Landlord's costing	Tenant's costing
9	4(7)	Computer trunking skirting to perimeter. Not laid against the floor leaving uneven gap.	Take up. Refit at lower level to close up the gap; make good finishes; redecorate.	A nominal gap between the floor slab and underside of the new electrical skirting trunking is noted. The new skirting trunking installed at the time of the reinstatement works has been lined through with the head of the original skirting trunking in situ to provide an even finish to the floor plate. Electrical skirting trunking is operable and has been provided with test certification. Therefore, no breach of covenant or disrepair has been identified by the landlord. Allow for providing cloaking pieces to base of new infill sections of skirting trunking.	Agreed, but cost would be £750.	2,750.00	2,000.00	750.00	500.00
10	4(7)	Ceiling tiles: damp staining already showing from roof leaks above. 2 No. tiles missing.	Replace missing and damaged tiles.	Item agreed. Cost excessive.	£100 reasonable.	100.00	20.00	100.00	20.00
11	4(7)	Smokers room window frames heavily stained from nicotine etc.	Replace cover frames left by tenant when this area was formerly a smoking room.	Location not identified – landlord to provide further evidence as to location and breach of covenant or disrepair identified	To follow.	950.00	n/a	n/a	n/a
12	4(7)	Plasterwork is damp damaged particularly at high level in lobbies adjacent to curtain wall enclosure.	Hack off damp damaged plasterwork and make good. Repair wall surfaces and prepare for redecoration.	Areas of damp and rainwater penetration have been recorded within the main stairwell at high level around the junction of curtain walling and flat roof soffit. We anticipate that there is an approximate 5m² in total patch repairs to be undertaken to the plaster finishes throughout this area. The landlord's claim and costs are excessive.	The area in question is more like 12m².	3,685.00	2,950.00	1,500.00	1,500.00
13	4(7)	Decorative finishes to wall surfaces have been damaged and are soiled.	Apply mist coat and two full coats emulsion paint finish to all previously decorated surfaces.	Tenant is under an obligation in the last year of the term to redecorate.	Agreed.	2,025.00	2,025.00	2,025.00	2,025.00

Section No 1: page 4 of 4

1	2	3	4	5	6	7	8	9	10
						Landlord's item		Tenant's item	
Item no	Clause no	Breach complained of	Remedial works required	Tenant's comments	Landlord's comments	Landlord's costing	Tenant's costing	Landlord's costing	Tenant's costing
				Escape stairs to west wing					
14	4(6)(b)	Plasterwork is damp damaged particularly at high level in lobbies adjacent to curtain wall enclosure.	Hack off damp damaged and cracked plasterwork and make good.	Approximately 20m² in total patch repairs of rainwater affected loose and off key plaster finishes are recorded. Landlord's claim excessive.	Area is about 60m².	3,685.00	800.00	1,200.00	800.00
15	4(6)(b)	There is extensive cracking to plastered surfaces.	Hack off damp damaged and cracked plasterwork and make good.	Landlord's claim and costs are excessive. Refer to item 14 above. Cost included in item 14.	See above.	Inc above	Inc above	Inc above	Inc above
16	4(6)(b)	Decorative finishes to wall surfaces have been damaged and are soiled.	Repair wall surfaces and prepare for redecoration. Apply mist coat and two full coats emulsion paint finish to all previously decorated surfaces.	Item agreed – tenant is under an obligation to redecorate within the last year of the term.	Agreed.	2,025.00	2,025.00	2,025.00	2,025.00
17	4(6)(b)	Ceiling tiles are generally discoloured. Many tiles are damaged, chipped or stained	Renew all suspended ceiling tiles, (replacement of damaged tiles only will result in a patchy and inconsistent appearance).	Refer to item 6.2.1. No disrepair accepted.	Disrepair is significant.	3,785.00	0.00	0.00	0.00
18	4(6)(b)	Carpets have widespread staining as a result of spillages and poor maintenance.	Take up and dispose of all carpet tiles. Supply and lay new carpet tiles of a grade and quality to match existing tiles.	Item agreed - carpet tile finish within westwing stairwell is aged and worn. Landlord's claim is excessive.	Agreed and reduced cost agreed	5,600.00	4,200.00	4,200.00	4,200.00
		TOTALS – REPAIR AND DECORATION				38,330.00	25,445.00	24,550.00	21,770.00

10/09/02

Appendix 5

Appendix 6

BIBLIOGRAPHY

Barnes, M. (ed), (1988) *Hill and Redman's Law of Landlord and Tenant*, 18th edition, Buttleworths, London, UK.

Dowding, N. and Reynolds, K. (2000) *Dilapidations: the Modern Law and Practice*, 2nd edition, Sweet and Maxwell, London, UK.

Lewison, K. (ed), (1978) *Woodfall: The Law of Landlord and Tenant*, 28th edition (amended and updated), Sweet and Maxwell, London, UK.

Smith, P. and West, W. (2001) *West and Smith: Law of Dilapidations*, 11th edition, Estates Gazette.

Williams, D. (ed), (1992) *Handbook of Dilapidations*, Sweet and Maxwell, London, UK.

Appendix 7

STATUTORY MATERIAL

Law of Property Act 1925: s146;

Landlord and Tenant Act 1927: s18;

Leasehold Property (Repairs) Act 1938;

Landlord and Tenant Act 1954: s51.

LAW OF PROPERTY ACT 1925

s146

1) A right of re-entry or forfeiture under any provisio or stipulation in a lease for a breach of any covenant or condition in the lease shall not be enforceable, by action or otherwise, unless and until the lessor serves on the lessee a notice–

   a) specifying the particular breach complained of; and

   b) if the breach is capable of remedy, requiring the lessee to remedy the breach; and

   c) in any case, requiring the lessee to make compensation in money for the breach;

   and the lessee fails, within a reasonable time thereafter, to remedy the breach, if it is capable of remedy, and to make reasonable compensation in money, to the satisfaction of the lessor, for the breach.

2) Where a lessor is proceeding, by action or otherwise, to enforce such a right of re-entry or forfeiture, the lessee may, in the lessor's action, if any, or in any action brought by himself, apply to the court for relief; and the court may grant or refuse relief, as the court, having regard to the proceedings and conduct of the parties under the foregoing provisions of this section, and to all the other circumstances, thinks fit; and in case of relief may grant it on such terms, if any, as to costs, expenses, damages, compensation, penalty, or otherwise, including the granting of an injunction to restrain any like breach in the future, as the court, in the circumstances of each case, thinks fit.

3) A lessor shall be entitled to recover as a debt due to him from a lessee, and in addition to damages (if any), all reasonable costs and expenses properly incurred by the lessor in the employment of a solicitor and surveyor or valuer, or otherwise, in reference to any breach giving rise to a right of re-entry or forfeiture which, at the request of the lessee, is waived by the lessor, or from which the lessee is relieved, under the provisions of this Act.

4) Where a lessor is proceeding by action or otherwise to enforce a right of re-entry or forfeiture under any covenant, proviso, or stipulation in a lease, or for non-payment of rent, the court may, on application by any person claiming as under-lessee any estate or interest in the property comprised in the lease or any part thereof, either in the lessor's action (if any) or in any action brought by such person for that purpose, make an order vesting, for the whole term of the lease or any less term, the property comprised in the lease or any part thereof in any person entitled as under-lessee to any estate or interest in such property upon such conditions as to execution of any deed or other document, payment of rent, costs, expenses, damages, compensation, giving security, or otherwise, as the court in the circumstances of each case may think fit, but in no case shall any such under-lessee be entitled to require a lease to be granted to him for any longer term than he had under his original sub-lease.

5) For the purposes of this section–

   a) "Lease" includes an original or derivative under-lease; also an agreement for a lease where the lessee has become entitled to have his lease granted; also a grant at a fee farm rent, or securing a rent by condition;

b) "Lessee" includes an original or derivative under-lessee, and the persons deriving title under a lessee; also a grantee under any such grant as aforesaid and the persons deriving title under him;

c) "Lessor" includes an original or derivative under-lessor, and the persons deriving title under a lessor; also a person making such grant as aforesaid and the persons deriving title under him;

d) "Under-lease" includes an agreement for an underlease where the under-lessee has become entitled to have his under lease granted;

e) "Under-lessee" includes any person deriving title under an under-lessee.

6) This section applies although the proviso or stipulation under which the right of re-entry or forfeiture accrues is inserted in the lease in pursuance of the directions of any Act of Parliament.

7) For the purposes of this section a lease limited to continue as long only as the lessee abstains from committing a breach of covenant shall be and take effect as a lease to continue for any longer term for which it could subsist, but determinable by a proviso for re-entry on such a breach.

8) This section does not extend–

i) To a covenant or condition against assigning, underletting, parting with the possession, or disposing of the land leased where the breach occurred before the commencement of this Act; or

ii) In the case of a mining lease, to a covenant or condition for allowing the lessor to have access to or inspect books, accounts, records, weighing machines or other things, or to enter or inspect the mine or the workings thereof.

9) This section does not apply to a condition for forfeiture on the bankruptcy of the lessee or on taking in execution of the lessee's interest if contained in a lease of–

a) Agricultural or pastoral land;

b) Mines or minerals;

c) A house used or intended to be used as a public-house or beershop;

d) A house let as a dwelling-house, with the use of any furniture, books, works of art, or other chattels not being in the nature of fixtures;

e) Any property with respect to which the personal qualifications of the tenant are of importance for the preservation of the value or character of the property, or on the ground of neighbourhood to the lessor, or to any person holding under him.

10) Where a condition of forfeiture on the bankruptcy of the lessee or on taking in execution of the lessee's interest is contained in any lease, other than a lease of any of the classes mentioned in the last subsection, then–

a) if the lessee's interest is sold within one year from the bankruptcy or taking in execution, this section applies to the forfeiture condition aforesaid;

b) if the lessee's interest is not sold before the expiration of that year, this section only applies to the forfeiture condition aforesaid during the first year from the date of the bankruptcy or taking in execution.

11) This section does not, save as otherwise mentioned, affect the law relating to re-entry or forfeiture or relief in case of non-payment of rent.

12) This section has effect notwithstanding any stipulation to the contrary.

LANDLORD AND TENANT ACT 1927

Part II

General Amendments of the Law of Landlord and Tenant

18 Provisions as to covenants to repair:

1) Damages for a breach of a covenant or agreement to keep or put premises in repair during the currency of a lease, or to leave or put premises in repair at the termination of a lease, whether such covenant or agreement is expressed or implied, and whether general or specific, shall in no case exceed the amount (if any) by which the value of the reversion (whether immediate or not) in the premises is diminished owing to the breach of such covenant or agreement as previously mentioned; and in particular no damage shall be recovered for a breach of any such covenant or agreement to leave or put premises in repair at the termination of a lease, if it is shown that the premises, in whatever state of repair they might be, would at or shortly after the termination of the tenancy have been or be pulled down, or such structural alterations made therein as would render valueless the repairs covered by the covenant or agreement.

2) A right of re-entry or forfeiture for a breach of any such covenant or agreement as aforesaid shall not be enforceable, by action or otherwise, unless the lessor proves that the fact that such a notice as is required by section one hundred and forty-six of the Law of Property Act, 1925, had been served on the lessee was known either–

    a) to the lessee; or

    b) to an under-lessee holding under an under-lease which reserved a nominal reversion only to the lessee; or

    c) to the person who last paid the rent due under the lease either on his own behalf or as agent for the lessee or under-lessee;

    and that a time reasonably sufficient to enable the repairs to be executed had elapsed since the time when the fact of the service of the notice came to the knowledge of any such person.

    Where a notice has been sent by registered post addressed to a person at his last known place of abode in the United Kingdom, then, for the purposes of this subsection, that person shall be deemed, unless the contrary is proved, to have had knowledge of the fact that the notice had been served as from the time at which the letter would have been delivered in the ordinary course of post.

    This subsection shall be construed as one with section one hundred and forty-six of the Law of Property Act, 1925.

3) This section applies whether the lease was created before or after the commencement of this Act.

LEASEHOLD PROPERTY (REPAIRS) ACT 1938

1 Restriction on enforcement of repairing covenants in long leases of small houses

1) Where a lessor serves on a lessee under subsection (1) of section one hundred and forty-six of the Law of Property Act 1925, a notice that relates to a breach of a covenant or agreement to keep or put in repair during the currency of the lease [all or any of the property comprised in the lease], and at the date of the service of the notice [three] years or more of the term of the lease remain unexpired, the lessee may within twenty-eight days from that date serve on the lessor a counter-notice to the effect that he claims the benefit of this Act.

2) A right to damages for a breach of such a covenant as aforesaid shall not be enforceable by action commenced at any time at which [three] years or more of the term of the lease remain unexpired unless the lessor has served on the lessee not less than one month before the commencement of the action such a notice as is specified in subsection (1) of section one hundred and forty-six of the Law of Property Act 1925, and where a notice is served under this subsection, the lessee may, within twenty-eight days from the date of the service thereof, serve on the lessor a counter-notice to the effect that he claims the benefit of this Act.

3) Where a counter-notice is served by a lessee under this section, then, notwithstanding anything in any enactment or rule of law, no proceedings, by action or otherwise, shall be taken by the lessor for the enforcement of any right of re-entry or forfeiture under any proviso or stipulation in the lease for breach of the covenant or agreement in question, or for damages for breach thereof, otherwise than with the leave of the court.

4) A notice served under subsection (1) of section one hundred and forty-six of the Law of Property Act 1925, in the circumstances specified in subsection (1) of this section, and a notice served under subsection (2) of this section shall not be valid unless it contains a statement, in characters not less conspicuous than those used in any other part of the notice, to the effect that the lessee is entitled under this Act to serve on the lessor a counter-notice claiming the benefit of this Act, and a statement in the like characters specifying the time within which, and the manner in which, under this Act a counter-notice may be served and specifying the name and address for service of the lessor.

5) Leave for the purposes of this section shall not be given unless the lessor proves:

   a) that the immediate remedying of the breach in question is requisite for preventing substantial diminution in the value of his reversion, or that the value thereof has been substantially diminished by the breach;

   b) that the immediate remedying of the breach is required for giving effect in relation to the [premises] to the purposes of any enactment, or of any byelaw or other provision having effect under an enactment, [or for giving effect to any order of a court or requirement of any authority under any enactment or any such byelaw or other provision as aforesaid];

- c) in a case in which the lessee is not in occupation of the whole of the [premises as respects which the covenant or agreement is proposed to be enforced], that the immediate remedying of the breach is required in the interests of the occupier of [those premises] or of part thereof;
- d) that the breach can be immediately remedied at an expense that is relatively small in comparison with the much greater expense that would probably be occasioned by postponement of the necessary work; or
- e) special circumstances which in the opinion of the court, render it just and equitable that leave should be given.

6) The court may, in granting or in refusing leave for the purposes of this section, impose such terms and conditions on the lessor or on the lessee as it may think fit.

2 Restriction on right to recover expenses of survey, etc.

A lessor on whom a counter-notice is served under the preceding section shall not be entitled to the benefit of subsection (3) of section one hundred and forty-six of the Law of Property Act 1925, (which relates to costs and expenses incurred by a lessor in reference to breaches of covenant), so far as regards any costs or expenses incurred in reference to the breach in question, unless he makes an application for leave for the purposes of the preceding section, and on such an application the court shall have power to direct whether and to what extent the lessor is to be entitled to the benefit thereof.

3 Saving for obligation to repair on taking possession

This Act shall not apply to a breach of a covenant or agreement in so far as it imposes on the lessee an obligation to put [premises] in repair that is to be performed upon the lessee taking possession of the premises or within a reasonable time thereafter.

4 [Repealed by the Landlord and Tenant Act 1954, s. 51(2)]

5 Application to past breaches

This Act applies to leases created, and to breaches occurring, before or after the commencement of this Act.

6 Court having jurisdiction under this Act

1) In this Act the expression "the court" means the county court, except in a case in which any proceedings by action for which leave may be given would have to be taken in a court other than the county court, and means in the said excepted case that other court.

2) ...

7 Application of certain provisions of 15 and 16 Geo. 5, c. 20

1) In this Act the expressions "lessor", and "lessee" and "lease" have the meanings assigned to them respectively by sections one hundred and forty-six and one hundred and fifty-four of the Law of Property Act 1925, except that they do not include any reference to such a grant as is mentioned in the said section one hundred and forty-six, or to the person making, or to the grantee under such a

grant, or to persons deriving title under such a person; and "lease" means a lease for a term of [seven years or more, not being a lease of an agricultural holding within the meaning of the [Agricultural Holdings Act 1986]] [which is a lease in relation to which that Act applies and not being a farm business tenancy within the meaning of the Agricultural Tenancies Act 1995].

2) The provisions of section one hundred and ninety-six of the said Act (which relate to the service of notices) shall extend to notices and counter-notices required or authorised by this Act.

8 Short title and extent

1) This Act may be cited as the Leasehold Property (Repairs) Act 1938.

2) This Act shall not extend to Scotland or to Northern Ireland.

LANDLORD AND TENANT ACT 1954

51 Extension of Leasehold Property (Repairs) Act 1938

1) The Leasehold Property (Repairs) Act, 1938 (which restricts the enforcement of repairing covenants in long leases of small houses) shall extend to every tenancy (whether of a house or of other property, and without regard to rateable value) where the following conditions are fulfilled, that is to say,–

   a) that the tenancy was granted for a term of years certain of not less than seven years;

   b) that three years or more of the term remain unexpired at the date of the service of the notice of dilapidations or, as the case may be, at the date of commencement of the action for damages; and

   c) that the [tenancy is neither a tenancy of an agricultural holding in relation to which the Agricultural Holdings Act 1986 applies nor a farm business tenancy.]

2) ...

3) The said Act of 1938 shall apply where there is an interest belonging to Her Majesty in right of the Crown or to a Government department, or held on behalf of Her Majesty for the purposes of a Government department, in like manner as if that interest were an interest not so belonging or held.

4) Subsection (2) of section twenty-three of the Landlord and Tenant Act, 1927 (which authorises a tenant to serve documents on the person to whom he has been paying rent) shall apply in relation to any counter-notice to be served under the said Act of 1938.

5) This section shall apply to tenancies granted, and to breaches occurring, before or after the commencement of this Act, except that it shall not apply where the notice of dilapidations was served, or the action for damages begun, before the commencement of this Act.

6) In this section the expression "notice of dilapidations" means a notice under subsection (1) of section one hundred and forty-six of the Law of Property Act, 1925.